Upstream Engineering and Flow Assurance Forum 2021

Held at the 2021 AIChE Annual Meeting

Boston, Massachusetts, USA and Online
7 - 11 November and 15 - 19 November 2021

ISBN: 978-1-7138-5718-1

Printed from e-media with permission by:

Curran Associates, Inc.
57 Morehouse Lane
Red Hook, NY 12571

CURRAN ASSOCIATES INC. proceedings .com

Some format issues inherent in the e-media version may also appear in this print version.

Printed with permission by Curran Associates, Inc. (2022)

For permission requests, please contact AIChE
at the address below.

AIChE
120 Wall Street, FL 23
New York, NY 10005-4020

Phone: (800) 242-4363
Fax: (203) 775-5177

www.aiche.org

Additional copies of this publication are available from:

Curran Associates, Inc.
57 Morehouse Lane
Red Hook, NY 12571 USA
Phone: 845-758-0400
Fax: 845-758-2633
Email: curran@proceedings.com
Web: www.proceedings.com

TABLE OF CONTENTS

630459 Development of an Electrochemical Approach to Detect Microbially Influenced Corrosion in Natural Gas Transmission Pipeline

Abstract:

This presentation describes a method for "detecting, mitigating and/or locating internal pipeline corrosion," through the development of an innovative electrochemical "signal" based approach to detect and quantify microbially influenced corrosion (MIC) in natural gas transmission pipelines. As natural gas production in the United States continues to increase, the abundance of transmission pipelines will increase similarly to accommodate the productivity. These pipelines are susceptible to internal MIC, which is exceedingly difficult to detect. First, MIC is by nature a localized process, and localizing damage is nearly impossible with "bulk" pipeline interrogation techniques. Second, highly specialized techniques are required to reliably diagnose MIC. Third, detection of internal MIC requires removal of samples from the pipeline for analysis and therefore, downtime. Fourth, evidence of MIC may not be evident until the system is "infected" beyond recovery.

This work focuses on the development of an electrochemical approach that involves zero-resistance ammetry (ZRA) measurements along a pipeline that can be used to detect MIC. The ZRA approach could be deployed along entire pipelines and would sensitively detect a broad range of in situ microbiological processes. To develop ZRA as an MIC monitoring tool, we carried out split chamber ZRA (SC-ZRA) incubations that entailed deployment of two steel working electrodes (WE1 and WE2) in chambers separated by a semipermeable membrane. Subsequently, one chamber is inoculated with an individual or consortia of microorganisms, and ZRA and potential measurements were made in conjunction with evaluations of microbiological activities. Current direction and magnitude can be indicative of the mechanisms and extents of MIC. The SC-ZRA setup mimics the heterogenous biofilm coverage of metal surfaces that leads to MIC, and we propose to use it to establish the electrochemical signatures of MIC. We used consortia and pure cultures of microorganisms from samples obtained from natural gas transmission pipelines to inoculate SC-ZRA and incubate them under conditions representative of pipeline fluids. We then specifically evaluated the impact of differential aeration and type of microbial metabolism (fermentation, sulfate reduction, methanogenesis, and acetogenesis) on current magnitude, direction, and extent of corrosion. The identification of microbiological processes with electrochemical signatures can then be used to inform the implementation of ZRA measurements as a robust, sensitive and broadly applicable MIC monitoring approach

630973 A Hybrid $CO_2/O_2/H_2s$ Corrosion Modeling Approach Using Machine Learning and CFD Velocity Data for General Pipe Configurations

Abstract:

Machine learning (ML) of CFD predicted CO_2 corrosion model have been demonstrated in our earlier publications using CFD as the baseline tool to predict the maximum CO_2 corrosion rates in pipes with bends [1][2][3]. However, it was found that further inclusion of other corrosion mechanisms requires a very large amount of CFD simulations deemed unfeasible in costs and time due to the coupling of $CO_2/O_2/H_2S$ corrosion kinetics [4], which makes the required number of parameters increased from 7 to 9. The increase does not seem significant but the required amount of the CFD simulations grows exponentially.

However, a careful examination of the CFD data in a full 3-D analysis shows that most of the surface chemistry is only dependent of the bulk chemistry data, i.e., the average species concentrations. This is due to the fact the surface corrosion chemistry occurs only near the wall surface and cannot be resolved by the CFD mesh. In fact, in CFD analysis the corrosion rate was approximated by a combination of near-wall mass transfer limitations based on the location variation of velocity, and chemical species concentrations. It was found that while the velocity local variation remains significant, the local variations of the resolved bulk (i.e. free stream) chemical species concentrations are negligible, therefore, the true surface corrosion rates are only impacted by the mass transfer limitations in most scenarios.

Nonetheless, there are scenarios where the local concentration variations cannot be ignored. In a pipe configuration such scenarios only occur when the surface corrosion rate is in the same order of the local mass transfer rate. This only occurs when the velocity is vey slow and the mass transfer rate near the wall is insufficient to remove the chemical species concentration gradient created by the surface reaction thus allows the gradient to be smoothed out by molecular diffusion. However, such process can only exist in laminar flows and is of little interest for the corrosion applications discussed here where the highest Damkohler number is only around 0.1. With this observation, the ML model is only required to predict the local velocity data and then combined with the corrosion kinetics developed by Nesic et al. [4][5] This approach greatly simplifies the CFD data generation process and allows a hybrid $CO_2/O_2/H_2S$ ML model to be developed.

In addition to the benefits of accelerated CFD data generation, this approach also enables further optimization of the Sherwood number-based mass transfer model. By using reinforcement learning algorithms the constant and exponents in the Sherwood number ($C\mathbf{Re}^n\mathbf{Sc}^m$) can be further optimized for better accuracy. The findings will be summarized and reported.

References

[1] Kuochen Tsai, "Corrosion modeling using electrochemistry and computational fluid dynamics", 2017 AIChE Annual Meeting, Minneapolis, MN, Oct. 29 – Nov. 3, Paper No. 335b.

[2] Yang, Huihui , Lu, Ligang , and Kuochen Tsai. "Machine Learning Based Predictive Models for CO_2 Corrosion in Pipelines with Various Bending Angles." Paper presented at the SPE Annual Technical Conference and Exhibition, Virtual, October 2020. doi: https://doi.org/10.2118/201275-MS.

[3] Kuochen Tsai, Huihui Yang and Ligang Lu, "Acceleration of CFD based corrosion models using Machine Learning Algorithms", Paper 67a, Nov. 16-20, 2020 Virtual AIChE Annual Meeting, San Francisco, USA.

[4] S. Nesic, J. Postlethwaite, and S. Olsen. "An Electrochemical Model for Prediction of Corrosion of Mild Steel in Aqueous Carbon Dioxide Solutions". CORROSION (1996) 52 (4): 280–294. https://doi.org/10.5006/1.3293640.

[5] S. Nesic, H. Li, J. Huang and D. Sormaz, "An open source mechanistic model for CO2/H2S Corrosion of carbon steel", Paper No. 09572, NACE Corrosion Conference & Expo, 2009.

632576 Sand Management in Wellbore and Surface Facilities using HYSYS

Abstract:

Most of the oil and gas reserves are contained in weakly consolidated sandstone reservoirs. The sandstones around the boreholes undergo a gradual degradation due to the drilling operation, cyclic effects of shut-in and start-up, reservoir pressure depletion, and strength-weakening effect of water [1]. The high pressure gradient caused due to fluid flow leads to detachment of sand, and the sand particles get carried away to the wellbore with the fluid flow. The amount of sand produced can be few grams per cubic meter of reservoir fluid, causing only minor problems or so high that it results in erosion, or in some cases filling and blocking of the wellbore [1]. The sand produced from the wellbore can also damage the surface equipment such as pipelines and pumps.

In case of a high oil production rate, erosional damage can occur due to the increased impact velocity of sand on the pipelines and fittings. On the other hand, when the oil production rate is low, sand accumulation occurs in the wellbore as the fluid velocity is not high enough to carry the sand particles. In both scenarios, sand production is undesirable, and a complete sand control could be considered an absolute solution. However, a complete sand control could cost the oil and gas industry millions of dollars while also reducing the oil production rate. A better approach for dealing with sand production is to find an operability range for the oil production rate such that both sand accumulation and erosion are minimal. For this calculation, the sand transport characteristics have to be estimated at various hotspots at different operating conditions, which is time-intensive. Hence, a tool is required, such that provided with the operating conditions, it estimates the sand transport characteristics under different sand production rates for a given wellbore and surface facility.

This work focuses on building frameworks to estimate the sand transport probability in the wellbore, the extent of sand particle erosion in the wellbore and fittings, and the rate of sand accumulation in the surface facilities. The framework is implemented within HYSYS Spreadsheets for ease of use. A complete model of the wellbore and surface facilities is simulated in HYSYS. The fluid properties necessary for the calculations mentioned above are imported from the simulation to the HYSYS spreadsheets. The framework uses these properties to perform the calculations and display the results.

In slurry transport, the critical velocity is defined as the minimum velocity in which the solids form a bed at the bottom of the pipe from fully suspended flows [3]. In the wellbore, when the fluid velocity is lower than the critical velocity for sand transport, sand gets deposited in the pipeline. As the deposition increases, the cross-sectional area of flow decreases, which leads to an increase in local fluid velocity and increased erosional damage. Hence it is crucial to ensure that the fluid velocity is greater than the critical velocity for sand transport. In the framework, the sand transport probability (STP) in the wellbore is estimated as the overlapping coefficient [2] between the normally distributed critical velocity of sand transport and the normally distributed actual velocity of the fluid in the pipeline. The critical velocity of sand transport in horizontal pipelines is calculated using the Oroskar and Turian model (1980), while that in the vertical pipelines is calculated using the Stokes' law. For inclined pipelines, the critical velocity is estimated by considering the critical velocities for the horizontal and vertical cases as the horizontal and vertical velocity components and interpolating between them for the angle of inclination of the pipe.

The framework consists of erosion response models from DNVGL-RP-0501 to estimate the extent of sand particle erosion in the pipelines and fittings. In these empirical models, the amount of material loss from a surface depends on the characteristic impact angles, impact velocity, sand properties, geometry, and target area [4].

In the surface equipment, sand is separated from the fluid by letting it settle in the equipment and then removing it. Sand is carried away by the fluid when the settling time of sand is more than the residence time of fluid in the equipment. The framework consists of a Residence Time Distribution (RTD) model to estimate the fraction of incoming fluid with a longer residence time than the settling time of sand to determine the fraction of sand that has settled. Here, the settling time of the sand is calculated using Stoke's law. The rate of sand settling and the rate of increase of sand bed height are also calculated.

The developed frameworks in the HYSYS Spreadsheets calculate and display the STP, extent of erosion, and rate of sand accumulation using the fluid properties imported from HYSYS Simulation to the spreadsheets. HYSYS Case Studies have been used to generate and plot the sand transport probability and extent of erosion results for a range of sand production rates using the frameworks. These results are used to determine the operability range for oil production with minimum sand accumulation and erosion. The rate of sand settling calculated for the surface equipment aids in deciding the operating conditions that ensure a high separation of sand from the fluid. The rate of increase of sand bed height calculated helps to monitor the sand build-up inside the surface facilities. This tool is used to carry out sensitivity studies of different parameters like sand properties, the volumetric fraction of sand, and different operating conditions on STP in pipelines, extent of erosion in fittings, and rate of accumulation of sand in surface facilities. It allows testing the performance of different proposed wellbore geometries in terms of sand transport characteristics to plan a new wellbore.

The tool was used to estimate the STP and erosion extent in a model wellbore. The wellbore consists of a series of horizontal or inclined pipe segments from the bottomhole, a vertical pipe segment that takes the oil to the surface, an elbow joint to connect the pipe segments, a tee-joint to represent the tree, and a choke on the surface. The estimated STP was higher for inclined and vertical pipes compared to horizontal pipes. The extent of erosion was the highest in the elbow. For an oil and water mixture having a flow rate of 22200 kg/h and 50% water-cut at 74 °C and 5400 psia carrying 10 pptb sand, the STP in a horizontal pipe decreased from 0.99 to 0.98, and the rate of erosion in the elbow increased from 3.77 nm/year to 6.29 nm/year with increasing the sand diameter from 60 microns to 100 microns. On the surface, for a vertical separator of 1.2 m diameter having the maximum level of liquid in the separator as 50% and the retention time as 5 min, the rate of sand accumulation increased from 3.45 mm/day to 7.79 mm/day with increasing the sand diameter.

References

[1] Rahmati, H., Jafarpour, M., Azadbakht, S., Nouri, A., Vaziri, H., Chan, D., & Xiao, Y. (2013). Review of sand production prediction models. *Journal of Petroleum Engineering, 2013*.

[2] Henry F. Inman & Edwin L. Bradley Jr (1989) The overlapping coefficient as a measure of agreement between probability distributions and point estimation of the overlap of two normal densities, Communications in Statistics - Theory and Methods, 18:10, 3851-3874.

[3] Oroskar, A. R., & Turian, R. M. (1980). The critical velocity in pipeline flow of slurries. *AIChE Journal, 26*(4), 550-558.

[4] DNV, G. (2015). Managing sand production and erosion. *Recommended Practice DNVGL-RP-O501. DNV GL Company, Oslo, Norway.*

620283 Field Wide Optimisation Towards Improved Hydrocarbon Recovery

Abstract:

Integrated Asset Modelling (also known as Integrated Production System Modelling - IPSM) is the modelling and simulation of an entire production facility consisting of both subsurface and surface elements. The intent is to holistically capture the complex interactions between the individual components of the system.

Historically asset modelling has entailed discrete reservoir, wells and facility models; however this mindset is slowly altering towards integrated asset modelling for production forecasts and optimisation whereby both system imposed constraints i.e. fluid flow from the reservoir and surface handling constraints can be considered simultaneously. This systems based approach is critical and reflects the current industry mind set. The full benefits of the integrated simulation tools employed within industry are not fully realised due to a number of
limitations; complexity of the reservoir models and hence limited usability in day to day production optimisation, overcoming the 'silo' discipline mentality that is common within interdisciplinary surveillance teams, simplification of integrated models is often in tension with accuracy and lastly, the limited capability of current tools to apply multi objective optimization analysis.

With these challenges, it can often be difficult to justify the effort and investment required to develop and deploy an integrated asset model. Despite these practical challenges, integrated asset modelling, if executed correctly can provide one of the best means of optimising production throughout field life with direct realized benefits such as extended field life and sustained production uptime.

The main objective of this research study is to understand the opportunities and limitations with the current existing industry integrated packages and develop a viable, representative alternative that is applicable for day to day surveillance and medium to long term field production optimisation. This research also aims to formally outline the drivers for integrated asset modelling followed by a review of the most common integrated asset modelling tools on the market in addition to review of the optimisation and reservoir modelling simplification methods available in literature. An integrated asset field model was developed including secondary recovery water injection as a test case.

Proxy models have been developed using proprietary HDMR meta modelling software at Imperial College London which was extended to the use of full physics reservoir models and associated sensitivity analysis.

These surrogate models have also been used in further heuristic based optimization frameworks and sensitivity analysis with very positive results in comparison to the field wide simulators thereby presenting a viable alternative offering significant advantages in simulation time.

623922 Effective Inhibition of Unfavorable Hydrate Formation in Magnetic Nanoparticle (MNP)-Stabilized Pickering Emulsions

Abstract:

Undesirable hydrate formation at an offshore drilling could induce clogging in a subsea pipeline, which could result in a huge environmental and economic damage. To prevent the unfavorable hydrate formation, non-ionic surfactant (Span 20)-adsorbed magnetic nanoparticles (MNPs, Fe_3O_4) were introduced. Through UV-Vis spectroscopy and SEM analysis, it was confirmed that the surfactant-adsorbed MNPs were effectively stabilized w/o Pickering emulsions. The Pickering emulsion stabilized by 0.125 wt% of MNPs were exhibited maximum ability of inhibiting hydrate formation, by blocking contact between methane molecules and water. Also, these MNP-stabilized emulsions lowered relative torque after hydrate formation, indicating that MNP-stabilized emulsions could act as an anti-agglomerants (AA). The inhibiting ability of MNPs were declined with higher concentration, due to increase of effective surface area of water from the formation of water bridges. By vibrating sample magnetometer (VSM) analysis and recovery simulations, it was revealed that MNP-stabilized emulsions had superparamagnetic behavior and MNPs could be successfully recovered by boot-drum separator.[1] Therefore, it can be concluded that surfactant-adsorbed MNPs could be utilized as a low-dosage hydrate inhibitor (LDHI), by acting as both kinetic hydrate inhibitor (KHI) and anti-agglomerants (AA).

1. J. Min, D.W. Kang, Y.H. Ahn, W. Lee, M. Cha, J.W. Lee, Recoverable magnetic nanoparticles as hydrate inhibitors, Chem. Eng. J. 389 (2020), 124461.

624197 Modeling of Wax Deposition in Production Pipelines

Abstract:

Wax deposition in production pipelines is a topic of concern in the oil industry. Wax gets crystallized when the temperature of the crude oil containing paraffin wax drops below the wax appearance temperature (WAT). Crystals may then accumulate on the surface of the wall leading to reduced production rate and higher pressure drop. Managing this issue is expensive due to the downtime associated with treatment methods. Therefore, an accurate model to predict the rate of asphaltene deposition is needed to address this topic. Majority of existing models predict the deposition rate by using molecular diffusion.

In this study, a model is developed based on the size of wax particles: dissolved, precipitated, and aggregated. The concept of a critical particle size and sticking probability are also applied in this model. The growth of wax particles over time is simulated using a population balance model, and particles bigger than the critical size are assumed to not be able to deposit due to the advection force of the flow. Experimental data points are used to verify the model performance.

622273 Effects of Shear and Pressure on Wax Appearance

Abstract:

Wax deposition in crude oil pipelines is a challenging issue for the oil industry. Paraffin wax deposition can result in pipeline blockage, which results in safety concerns and financial losses. Factors affecting wax deposition include wax concentration, cooling rate, pressure, and flow. Despite the importance of considering shear rate on equipment and pipeline design, how shear rate may alter wax appearance temperature (WAT) and gel strength has not been widely studied. At reservoir pressures light gases, such as methane, can dissolve in oil. Thus, wax appearance might be different than the phase transition observed at ambient pressure. We investigated samples of light mineral oil and paraffin wax at different wax concentrations. Temperature ramp experiments identified an increase in viscosity related to wax appearance under steady shear. Yield stress, on the other hand, was measured using oscillatory sweep experiments. A high pressure, concentric cylinder rheology cell was used to expand study beyond ambient pressure, parallel plate measurements. WAT increased significantly by increasing wax concentration. Changing concentration from 5 to 50 wt% resulted in a~ 20 °C increase in WAT at shear rate 1000 s^{-1} and also at static conditions. Viscosity after wax formation increased with concentration to the power of 1.5 at shear 10 s^{-1}. Lowering cooling rate from 10 to 0.2 °C /min under shear rate of 100 s^{-1} increased WAT up to 4 °C. Shear rate had a more modest effect on WAT of a ~ 2 °C decrease when shear rate increased from 1 to 1000 s^{-1}. However, shear rate has a significant effect on the yield stress of formed gels. An increase in shear rate from 0 to 1000 s^{-1} caused the yield stress to decrease from 415 to 8 Pa. WAT decreased up to 4 °C when pressure increased from ambient pressure to 1400 psi. Overall, adding shear rate and pressure to phase behavior studies could help the oil industry better understand wax flow assurance issues.

621902 Experimental Study on the Deposition Characteristics of Wellbore with Highly Resin and Asphaltene Crude Oil

Abstract:

Wax deposition is one of the core issues in the flow assurance study of oil and gas wellbores, particularly in deep and ultra-deep water conditions. The component properties of the crude oil vary widely in different seas, which further complicates wax deposition issue. Although recent advances have been made in understanding the performance and mechanism of wax deposition, the influence of asphaltene and resin, remains an open question.

In this work, the effects of asphaltene and resin on the wax deposition layer in J and B oil wellbores with the variation of different production rates were investigated by using the cold finger and HTGC. Compared to the deposition results of the two wellbores, the initial deposition layer of J oil is thicker and the increase of deposition thickness with time is smaller, and the deposition also occurs when the wall temperature is higher than the wax appearance temperature (see Figure 1A), and the distribution of carbon number of wellbore plug and oil sample is basically the same (see Figure 1C), so the aging effect is weaker. The gelling effect caused by the high asphaltene content of J oil cannot be ignored. However, the deposition characteristics of oil B are in accordance with the typical wax deposition law, i.e., the deposition thickness increases when the temperature decreases/the temperature difference increases (see Figure 1B), and the content of recombination (C40-C60) in the wellbore plug is significantly higher than that of the oil sample, while the content of lighter components (C7-C40) is lower than that of the oil sample, obeying the aging law.

This work adds to growing evidence of the coexistence of crude oil gelling and wax deposition. These conclusions also provide theoretical reference for the wax prevention and removal methods.

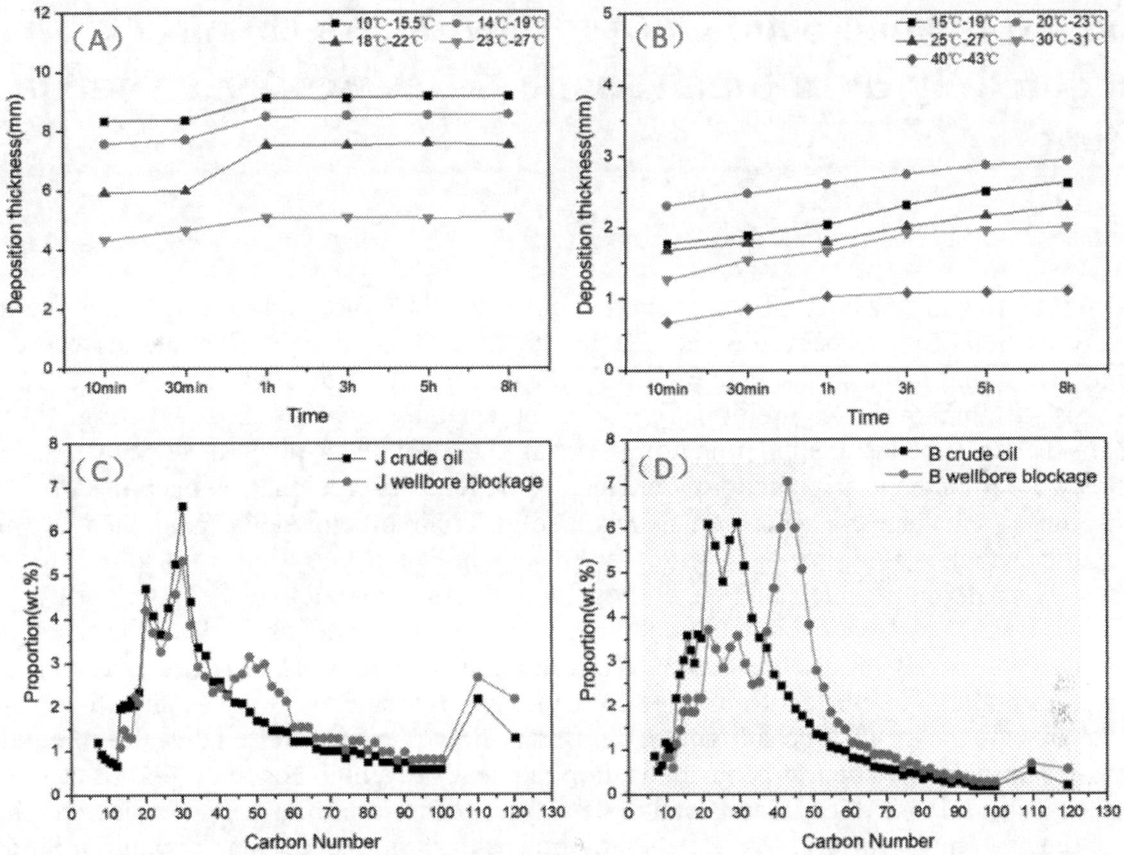

Figure 1. The thickness and carbon number distribution of the deposition layer in J&B wellbore

632766 Asphaltene-Solid Co-Aggregation Mechanism Investigated By Ultra-Small Angle X-Ray Scattering and *in Situ* Titration

Abstract:

Athabasca bitumen contains over 10% asphaltenes and around 1% inorganic solids following the froth extraction process. The previous results showed that the asphaltene precipitation rate increased significantly when inorganic solids were present in the mixture [1]. However, the co-aggregation mechanism is still unknown. Our preliminary results indicated there might exist two competitive mechanisms. Heterogeneous precipitation dominates at low insoluble asphaltene concentration, as more solids were found in precipitated asphaltenes. Homogeneous precipitation becomes the majority at high insoluble asphaltene concentration, as fewer solids are in precipitated asphaltenes. We aim to investigate the asphaltene-solid co-aggregation at low and high concentrations of insoluble asphaltenes. To profoundly understand the co-aggregation, ultra-small angle X-ray scattering (USAXS) was used to observe the interaction between nanometer-scale insoluble asphaltenes and solids. USAXS effectively probes the structural information, size, internal structure, overall shape, and surface morphology of clusters from 5 nm to 5 microns. Heptane was used as an anti-solvent and titrated into an asphaltenes-toluene solution at a different flow rate for precise control of the rate of asphaltene destabilization. The fast heptane titration rate leads to a high concentration of insoluble asphaltene in a short time, which is supposed to be an ideal condition for homogeneous precipitation in terms of the preliminary results. We are able to observe the homogeneous precipitation. Slowly heptane titration created an ideal condition for heterogeneous precipitation, propitious to exploring heterogeneous precipitation. This study has the potential to assistant us in controlling asphaltene-solid co-aggregation for more efficient removal of solids from bitumen during naphthenic froth treatment.

[1] Rapid Heterogeneous Asphaltene Precipitation with Dispersed Solids W Chaisoontornyotin, J Zhang, S Ng, MP Hoepfner - Energy & fuels, 2018

630255 An Economic Approach for Non-Recovered Oil Recovery: Microbial Enhanced Oil Recovery

Abstract:

These days, crude oil leads to the source and supply of the majority of the world's energy requirement. The reserves are available but the recovery of the oil through the conventional method always remains a question. The conventional method of recovery along with the secondary methods leave an enormous amount of crude left behind unrecovered in the reserve only.

There comes the need of tertiary method which is characterized as the enhanced oil recovery (EOR) for the recovery of the unrecovered. Among all the available EORs, Microbial Enhanced Oil Recovery (MEOR) has been proven to be a technologically advanced resource to uncover the non-renewable resources.

MEOR utilizes the biomeric material especially the microbes to alter the properties of oil and the supports the oil movement and enhances the recovery. The studies provided the implication of the MEOR with low cost and high outputs. The present study provides the exploitation of the diverse microbial systems such as bio-surfactants, bio-polymer, solvents and enzymes for amending the properties and enhancing the recovery of the non-recovered crude oil.

632994 Gas Production Strategies and the Impact of Layers on Gas Recovery from Gas Hydrate Reservoir Using the Numerical Simulations

Abstract:

The gas hydrates can be the efficient alternative to replace the depleting conventional energy production method. The energy demand of a growing country like India can be alleviated by extracting the extensive methane gas present in the gas hydrate reservoirs found at multiple sites. The possibility of the gas production form methane hydrate reservoir is evident from the short-term field tests; however, the reservoir complexities still challenge the long-term production. The prognostication of the methane production from the gas hydrate reservoirs is usually done by simulation studies to address the challenges in the long-term productions. Therefore, the production analysis requires the grasp of geological, technical, and economical factors controlling the methane production potential of these reservoirs. In the present work we analyze the efficient production method as well as the gas production behavior due to different well arrangements and the well locations. We also study the impact of layering on the percent gas recovery from the gas hydrate reservoir. A 3-D, class-2 oceanic reservoir block underlain by an infinite aquifer layer is modeled in this work.

An In-house, thermal 3-D finite volume legacy simulator is used with multicomponent- water, methane, and hydrate in four phases – gas, aqueous, hydrate and ice. Energy and mass balance equations are solved in space and time domain to compute the gas recovery from the gas hydrate reservoir. The unconfined reservoir becomes less responsive to the depressurization method for both the vertical as well as horizontal wells. Therefore, the warm water injector is a must along with the depressurization method to recover methane gas from an unconfined gas hydrate reservoir. The different well arrangements and well locations are explored using numerical simulations. The wells placed horizontally are more effective where gas recovery is 48% original gas in place (OGIP) compared to the gas recovery of 22% of OGIP for vertically placed wells. The location of the horizontal injector near the aquifer makes water convection easier and the gas recovery starts from the very beginning. The location of the injector in the low permeability hydrate layer makes the gas recovery difficult for initial days. We also investigate the impact of layering on the gas recovery from the gas hydrate reservoirs. The effective permeability of the topmost layer of hydrate where producer is located plays an important role in gas recovery. The infinite aquifer layer has impact on both the gas and the water recovery because the higher effective permeability of the water layer supports the pressure drop in the gas hydrate reservoir.

632639 Prediction of Production Performance of Unconventional Reservoirs Using Data-Driven Methods

Abstract:

Evaluation of the early production data plays a vital role in minimizing uncertainties associated with developing unconventional reservoirs. The traditional decline curve analyses are not reliable for analyzing the early production data due to the long transient flow in unconventional reservoirs. Other challenges of developing unconventional reservoirs include multiphase flow, uncertain geological properties, complex fluid flow mechanisms, and variable completion properties, which cannot be easily modeled using empirical decline models. Fundamental equations and reservoir simulations can also be used for modeling the production performance of unconventional reservoirs, which are effective but also time-consuming. Furthermore, it doesn't satisfy the current need to drill many wells in a short amount of time to improve the return on investment.

This study presents a data-driven approach to forecast the production from unconventional reservoirs with limited production histories using multivariate statistics and traditional decline curve analysis. We use an unsupervised machine learning algorithm to identify the patterns and regularities on a large production data set. This algorithm is then applied to large sets of field data to create an uncorrelated weighted sum of 'representative curves,' the truncated summation of which yields an approximation of the original data.

The method presented is also applicable to estimate the gas to oil ratio (GOR) for liquid-rich shale reservoirs with complex phase behavior. As the underlying data capture the effect of multiphase flow automatically, this method presents a reliable alternative to estimate the secondary phase from volatile oil and gas condensate reservoirs.

AUTHOR INDEX